© Aladdin Books Ltd

Design: Malcolm Smythe
Text: Kate Petty

ISBN 0 531 04903 5

LCCC No. 84 52509

Separation by La Cromolito, Milan
Typesetting by Dorchester Typesetting
Printed in Belgium

Designed and produced by
Aladdin Books Ltd
70 Old Compton Street
London W1

*First published in
the United States in 1985 by*
Franklin Watts
387 Park Avenue South
New York, NY10016

Contents

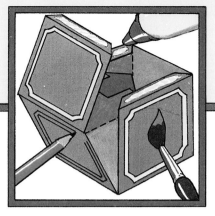

Build Your Own

SPACE STATION

Consultant Caroline Pitcher

Illustrated by Louise Nevett

Franklin Watts

New York · London · Toronto · Sydney

ABOUT THIS BOOK

The project in this book has been designed so that children can build it unaided, either individually or in groups.

Non-readers can follow the step-by-step pictorial instructions. A black key illustration of the finished project appears on each spread with the unit to be completed colored in red.

The materials for making the project are generally available at home and in classrooms, but alternatives are often suggested. The materials which you could use for making the project are illustrated opposite. It is a good idea to make a habit of collecting all sorts of household bits and pieces (see p.30). Always rinse out bottles and cartons and make sure they have not contained any dangerous liquids such as bleach or disinfectant.

Tracing

When an outline has to be traced, fix tracing paper over the outline. Trace the outline. Turn over the tracing and rub pencil on the back of the pencil line. Tape the tracing, outline upwards, on paper or cardboard and retrace outline.

Cutting

Make sure that the scissors which children use have rounded ends for safety and never give children a sharp knife. The point of a sharp pencil is a safe and effective way of making holes in cardboard.

Paper and Card

These can be distinguished in the instructions by the different colors shown here. Paper is white and card is blue in color. Remember to cut along solid lines and fold along dotted lines.

Glueing

Any sort of ordinary paste or glue is suitable for making the project but very strong glue can be dangerous and should be avoided.

Coloring

For shiny or plastic surfaces use poster paint or powder paint. Ordinary powder paint or watercolor can be used successfully on other surfaces. Alternative coloring methods are wax crayons, colored pencils or felt-tip pens and painting sticks. Large areas can be covered with colored paper. Make sure that the colors you use are lead-free and nontoxic.

WHAT YOU MAY NEED

Scissors

Cork

Scotch tape

Coins

Toothpicks

Rubber bands

Glue

Knitting yarn

Cotton

Modeling clay

Cotton thread spool

Tinfoil

Yogurt or cream container

Paper

Cardboard

Small box

Matchboxes

Used matchsticks

Kitchen knife

Pencil

Paintbrush

Pipe cleaner

Drinking straw

String

Plastic bottle

Quart milk or juice carton

One-pint milk or juice carton

Large cardboard roll

Small cardboard roll

Large box

5

SPACESHIP

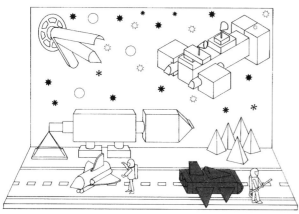

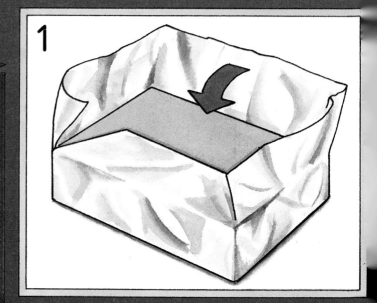

1

★ A small tissue box is a good size for the body of the spaceship, but the exact sizes of the parts in this project are not especially important, so long as they are in proportion.

★ Cover the box with tinfoil.

★ Use a candy tube or make your own tube from card. Shape the cockpit by cutting a slit about one-third of the way down. Overlap the ends and secure with scotch tape.

★ Use the body of the spaceship as a guide to the size of the wings. The wings and the tailplane can be made more secure in their slots with scotch tape.

★ Make the paper tubes for the jets so that they fit inside the box at the rear of the spaceship. Paint them before you glue them to the inside of the box.

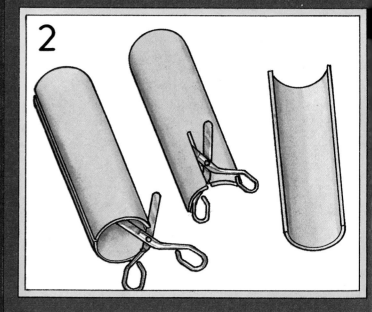

2

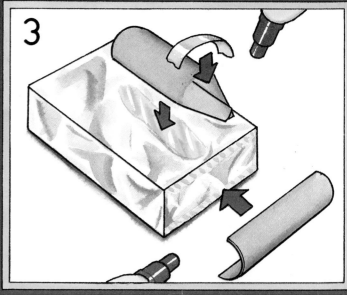

3

4

5

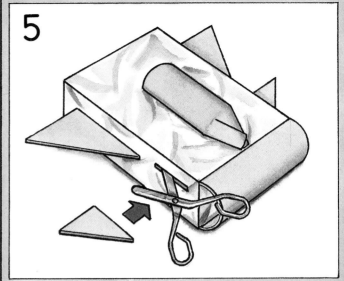

6

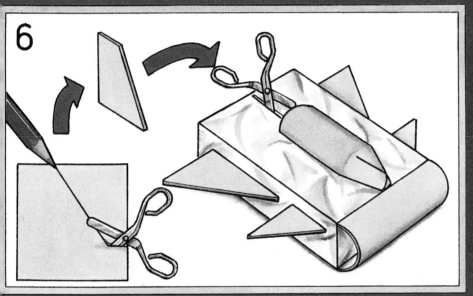

7

8

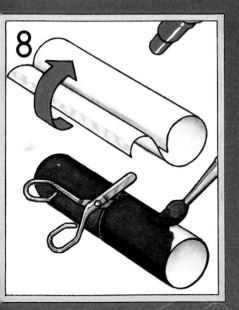

9

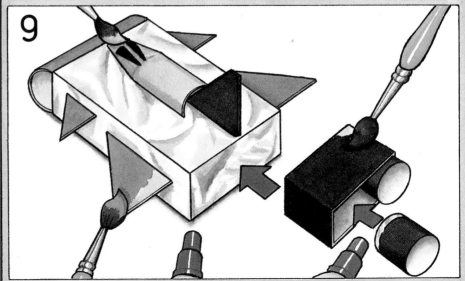

SPACE COLONY

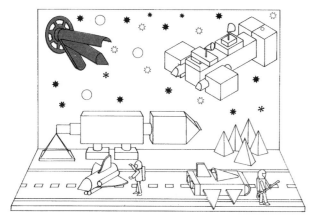

★ Cover a small plastic bottle with white paper. You might prefer to draw the pictures on the flat piece of paper first. If you do not have a bottle you can make a tube from rolled card. Think up different scenes to take place on each level of the colony.

★ Use the bottle to measure the card for the wings. The edges should meet rather than overlap.

★ Cut the wings down as far as the bottom of the lowest picture on the bottle. Secure them around the bottle with scotch tape.

★ The ring should fit over the top of the space colony. The glue will help it stay in place.

★ Decorate the wings of the space colony in any way you like. You could give it a name and stick the letters on.

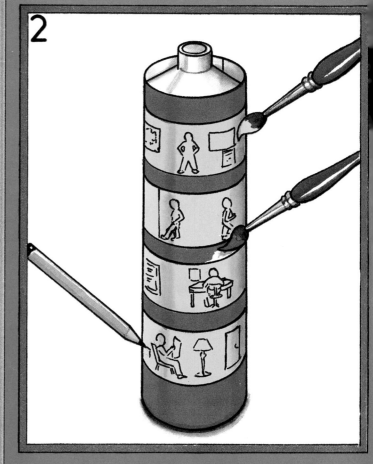

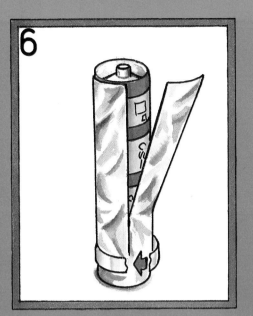

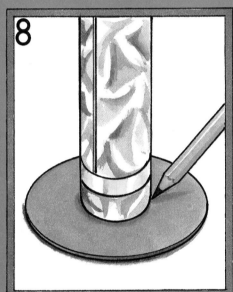

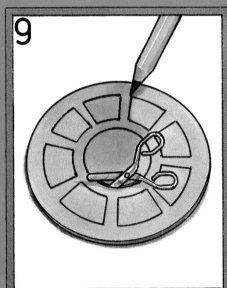

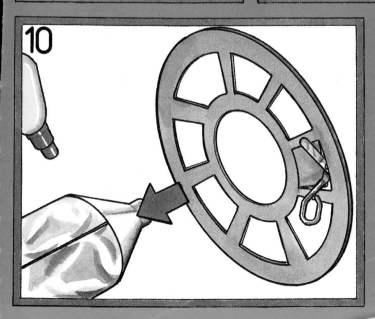

WALKING FACTORY 1

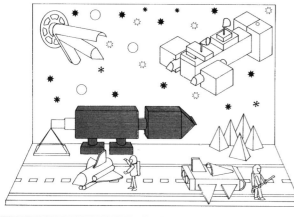

★ You need three cardboard rolls for this project. You can make them yourself by rolling and glueing card if necessary. The body of the factory is best made from a quart milk or juice carton and a one-pint one.

★ Open up the top of the quart carton carefully and cut down the corners. Use scotch tape to hold it in place around the cardboard tube. Seal the top of the one-pint carton.

★ The cardboard rolls should slot into the holes you cut for them, but a little glue will help secure them.

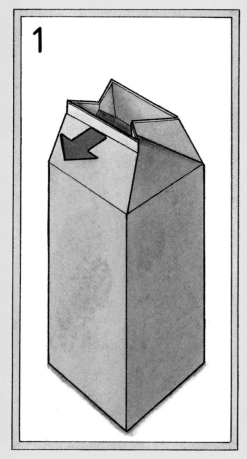

1

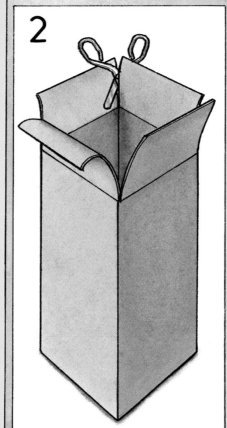

2

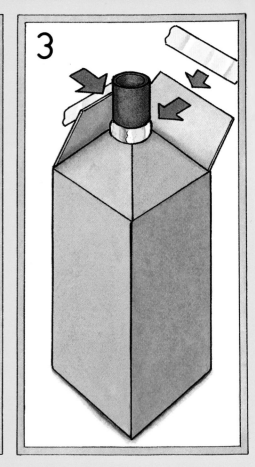

3

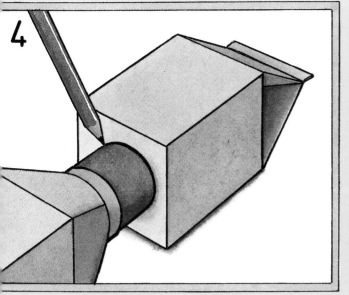

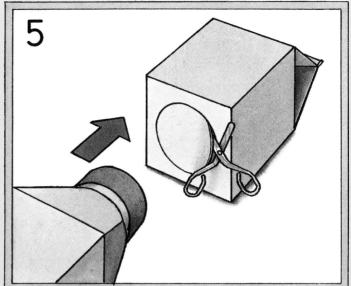

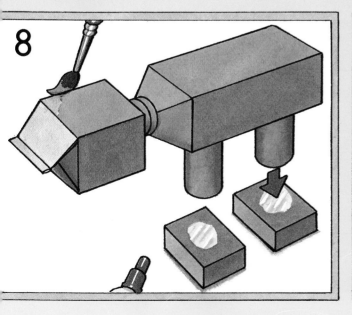

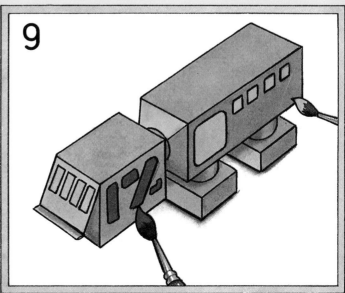

WALKING FACTORY 2

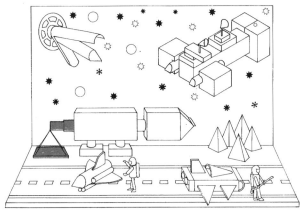

★ You will need two more cardboard rolls if you want to add a freight platform to your walking factory. One of them should be able to slide into the other one.

★ Use the point of a very sharp pencil to make holes in the smaller cardboard roll to thread the straw through. You could use a plastic stirrer instead.

★ Your four pieces of thread or string should be of equal length. Knot them into the holes and tie them together at the top.

★ Attach the freight platform to the winch so that it can be raised or lowered to the ground.

★ You can make freight and supplies to be carried on the platform from anything you like.

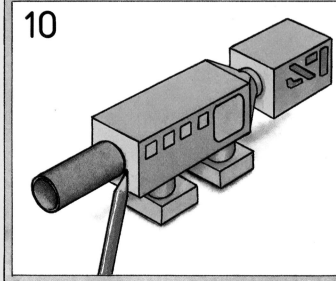

10

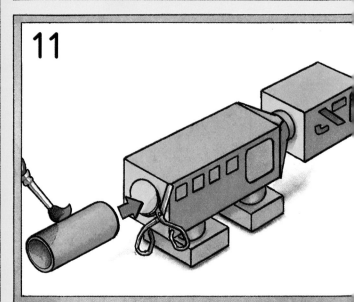

11

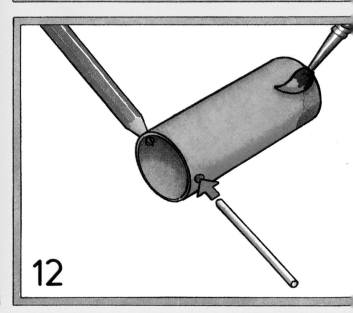

12

13

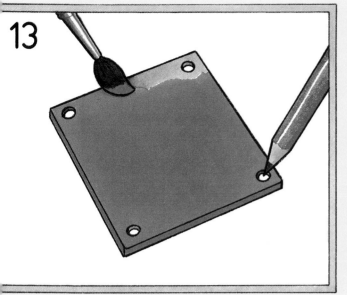

14

15

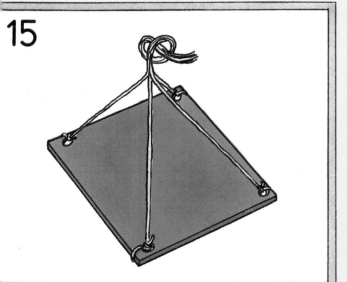

16

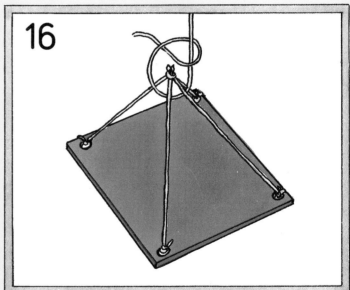

17

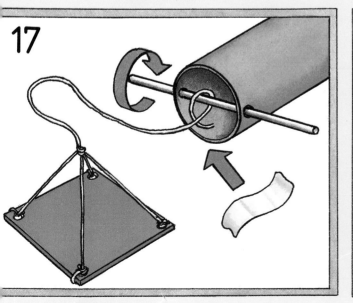

18

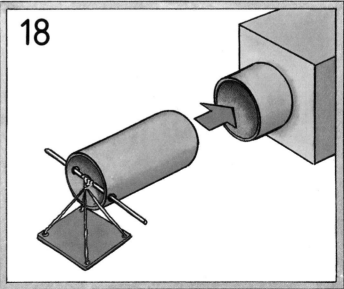

SPACE CRUISER 1: PARTS

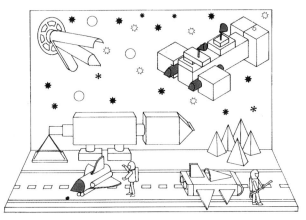

★ The pictures on this page show how to make the solar panels, nose cones, radar dishes, rocket boosters and jets which you need for your space cruiser.

★ The card for the solar panel should be about the same size as a matchbox. The model in the picture has one solar panel but you can give it as many as you like.

★ Make three nose cones and two radar dishes from circles of paper which have a single cut to the center. The nose cones will have to be fitted to the two rocket boosters shown here and to the shuttle on page 23.

★ Make the rocket boosters from thin card. Fit them round a cardboard tube, to make them exactly the same size.

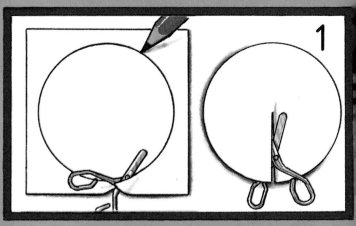

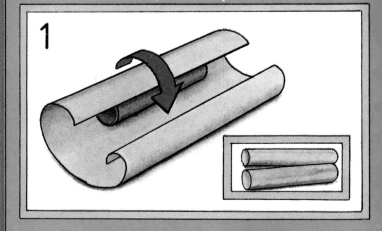

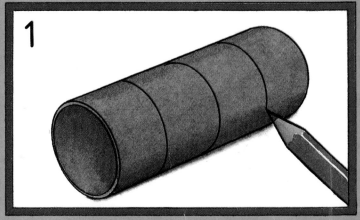

2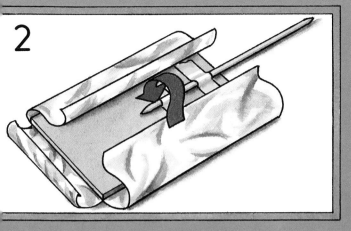

3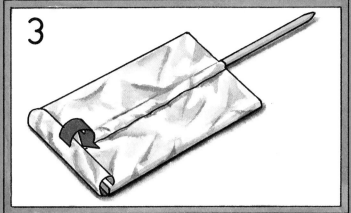

2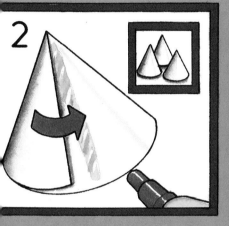

3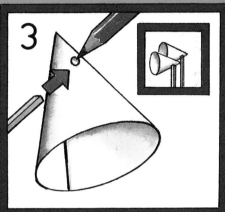

4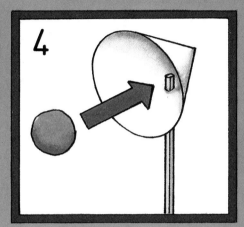

2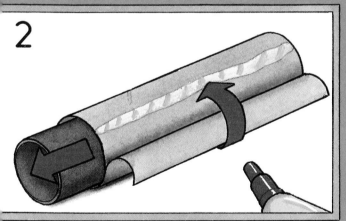

3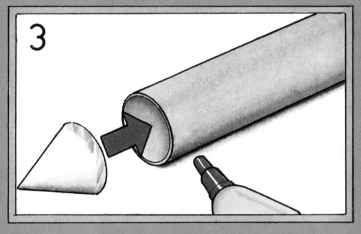

2

3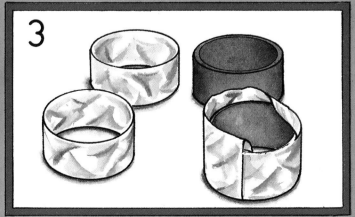

SPACE CRUISER 2

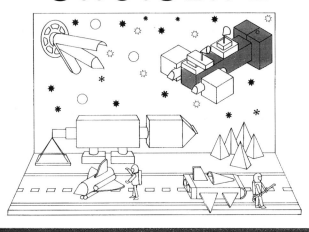

★ You have already been shown, on page 17, how to make the jets in pictures 7 and 8.

★ Join two quart milk or juice cartons to make the body of the space cruiser. You will need to open out the top of one of them and close the other.

★ The rear of the space cruiser is made from two one-pint cartons, but a single box of the same size will do instead. If you use cartons make the ends neat by cutting down the corners of the top and sticking down the flaps.

1

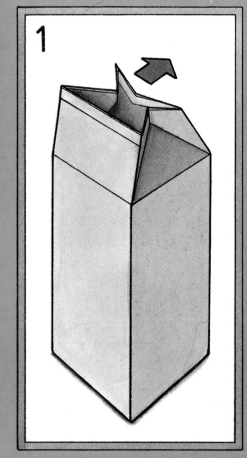

2

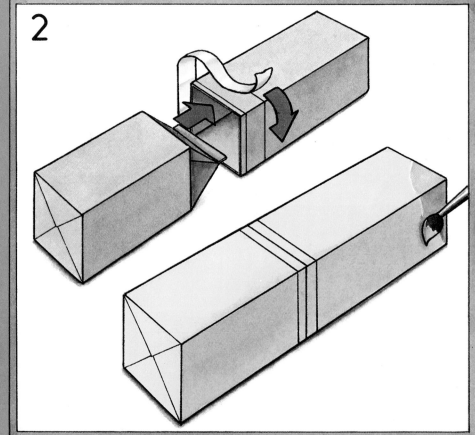

3

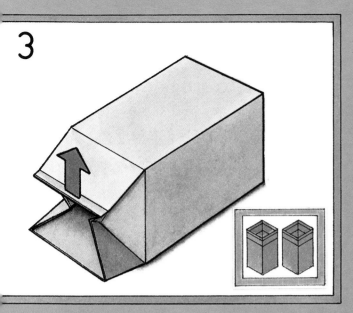

4

5

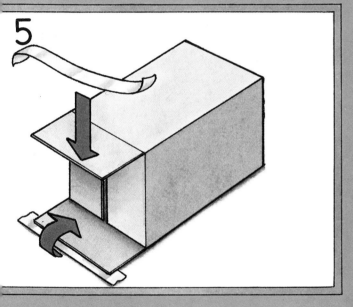

6

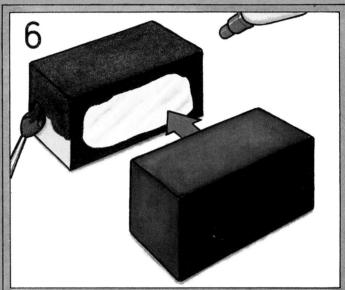

7

8

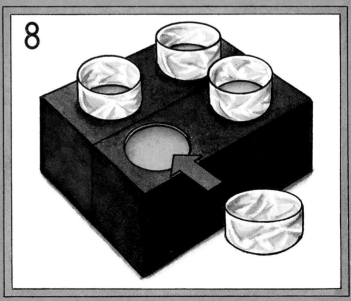

SPACE CRUISER 3

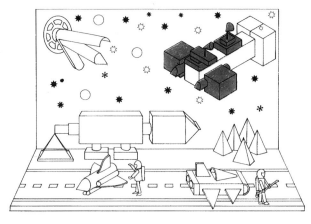

★ You have already been shown how to make the rocket boosters, radar dishes and a solar panel on page 17.

★ The boosters are mounted on two one-pint cartons, but similar sized boxes will do instead. If you use cartons make the ends neat as shown on the previous page.

★ Shape the grooves to fit the boosters carefully. Mount the boosters in the grooves and secure with scotch tape

★ Different sized matchboxes make good mountings for the solar panel and radar dishes. The pictures here give suggestions but you might like to make more panels and dishes and attach them any way you like.

★ Make sure the jets at the back and the boosters at the side are all going to propel the cruiser in the same direction.

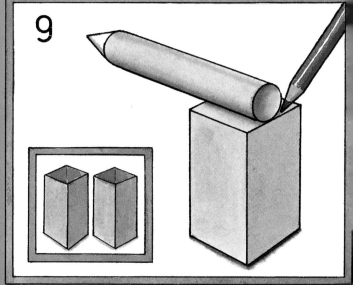

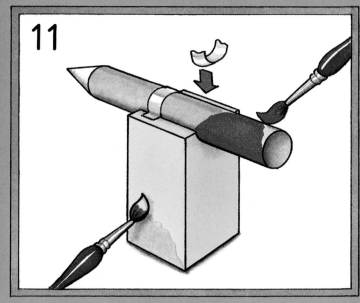

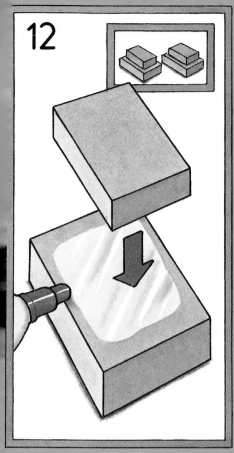

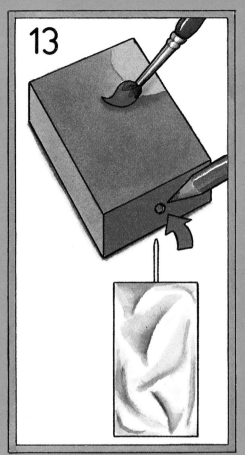

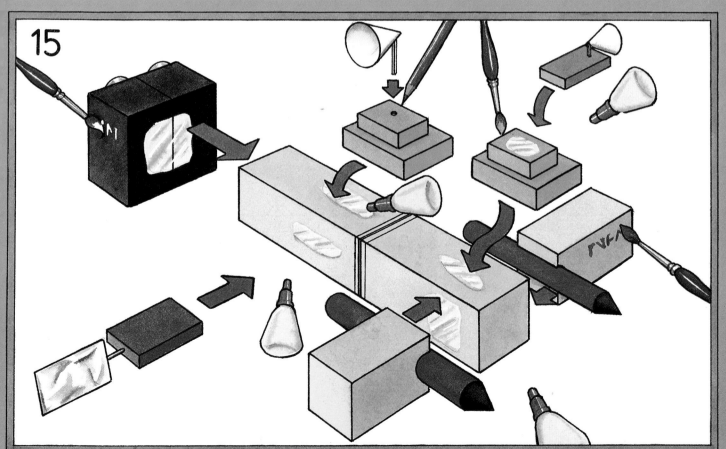

SPACE CRUISER SHUTTLE

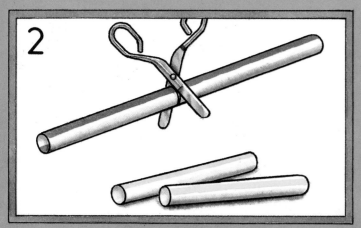

★ When you are making your cardboard tube for the body and drawing the wings, remember that the completed shuttle has to be small enough to fit inside the cruiser.

★ You have already been shown how to make the nose cone on page 17. Make sure the cone fits the shuttle before you glue it together.

★ Take care when you cut out the door of the shuttle bay. You must push the scissors through the carton and cut around the line. Remember that the door is hinged at the bottom.

1

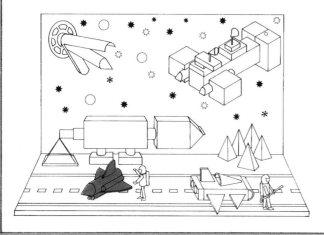

2

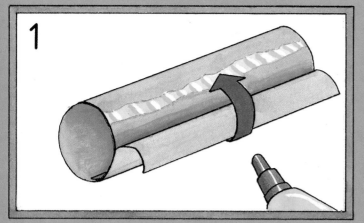

3

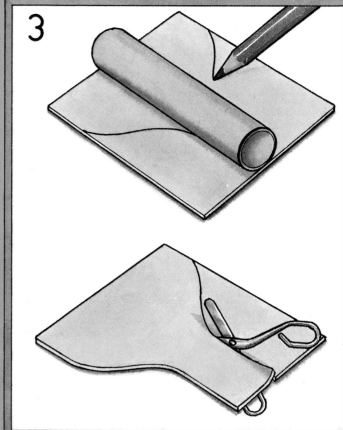

4

5

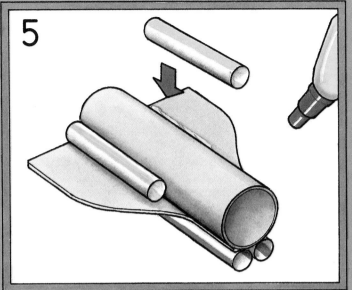

6

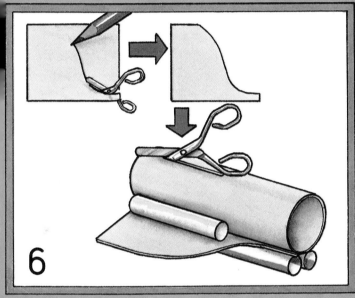

7

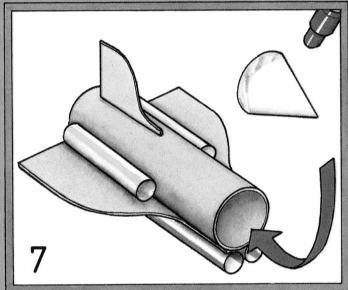

8

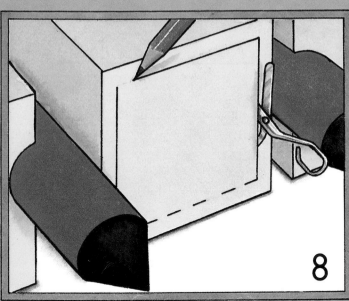

9

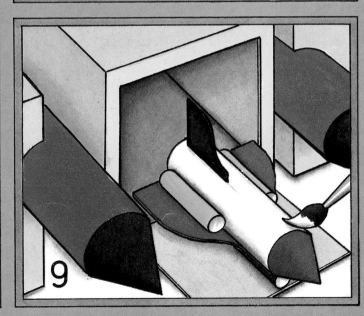

SPACE PEOPLE

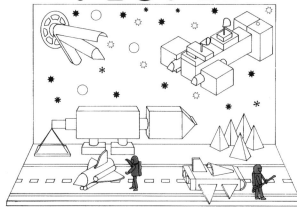

★ The pictures here show how to decorate hardboiled eggs as heads for space people, but you can use pebbles or modeling clay depending on the size you want your completed figures to be.

★ Use a sharp pencil to make holes in the body. Push the foil-covered straw through the holes and cut to size. If you thread a pipe cleaner through the straw you can bend the arms and also make hands.

★ Attach one end of the yarn lifeline under the oxygen pack. The other end can be connected to a vehicle on the ground.

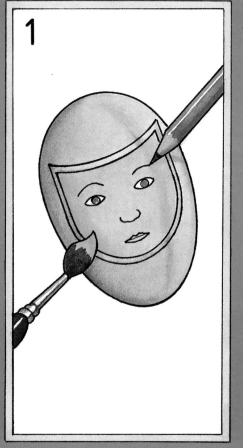

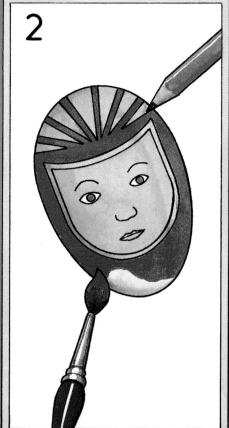

4

5

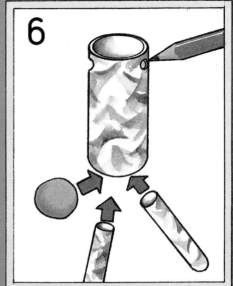

6

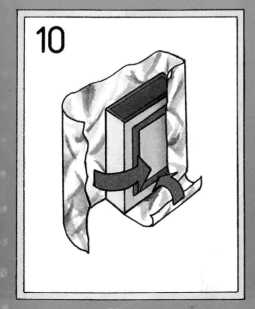

7

8

9

10

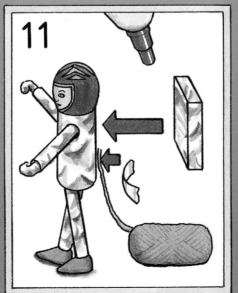

11

12

SPACE STATION BUILDINGS

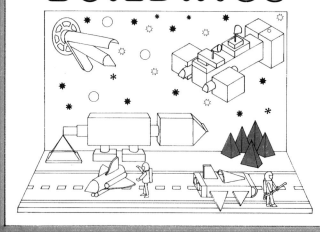

★ Trace the outlines of the buildings in picture 1 and picture 2 onto thin card.

★ The dotted lines are for folding and the solid lines show where to cut.

★ Cover the unmarked side of the shape in tinfoil. Fold the shape inwards so that the foil-covered side forms the outside of the shape.

★ Make as many buildings as you like. You can make different shapes by joining some of them together.

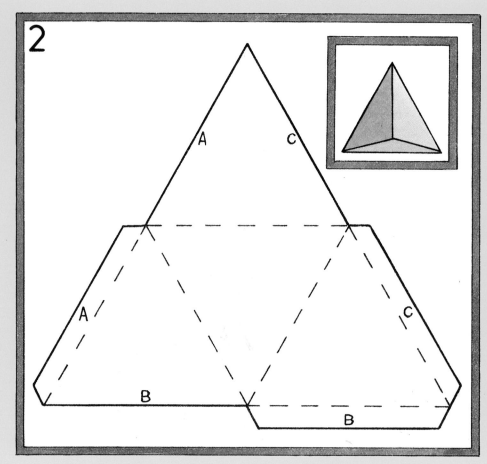

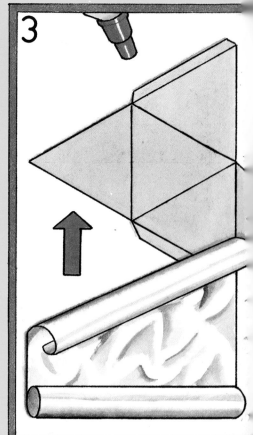

26

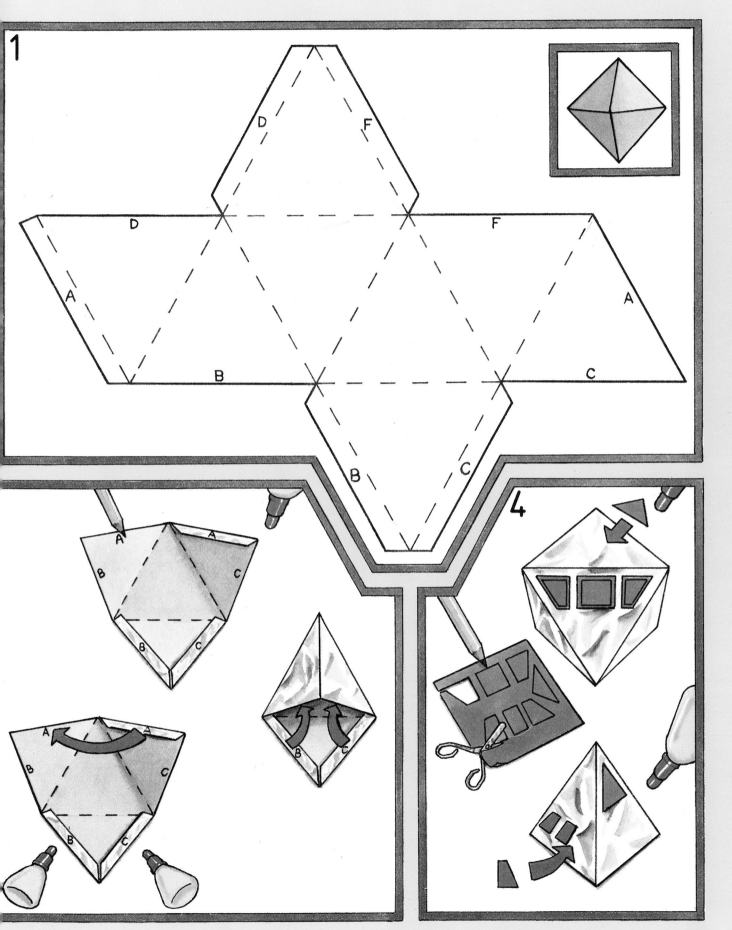

1

D F

D F

A A

B C

B C

4

A A

B C

B C

A A

B C

B C

B C

SPACE STATION ASSEMBLY

★ To display your space station most effectively, push a table against a wall and make a dark blue background as shown.

★ You can copy the shapes from the picture or make up your own to decorate the backdrop.

★ You might like to suspend the space cruiser and the space colony from the ceiling or from a shelf above the table.

★ Sand, gravel or pebbles make a good surface. You can make larger rocks from crumpled newspaper secured with scotch tape and painted.

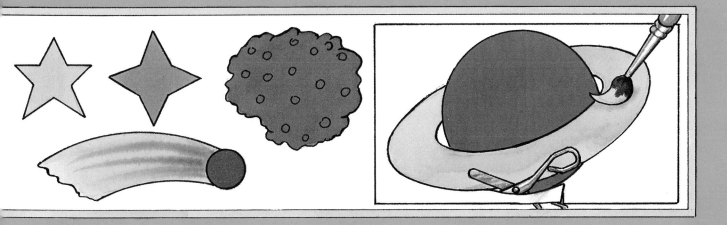

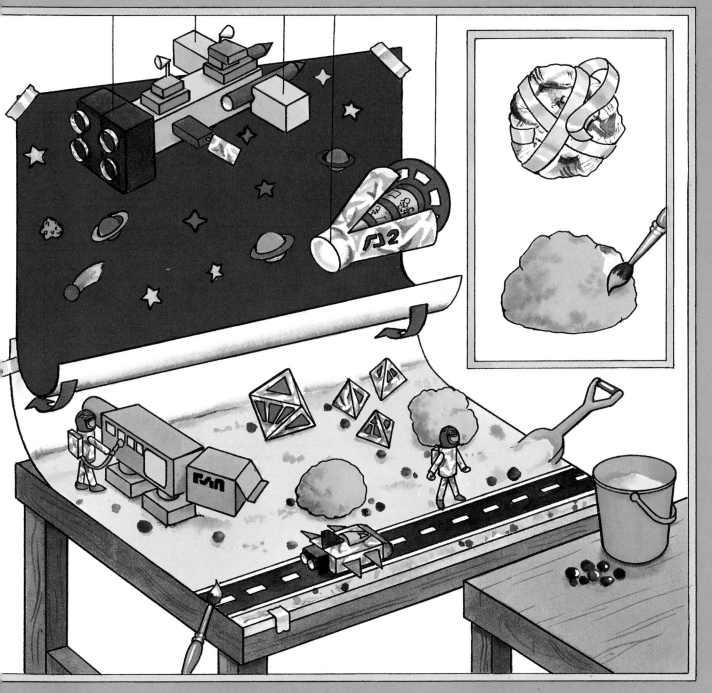

BITS BOX

Collect all sorts of household bits and pieces on a regular basis.
They can provide an endless source of creative play for children, not
only for making the projects in this book but also enabling children
to invent models of their own. Keep everything together in a "bits box."

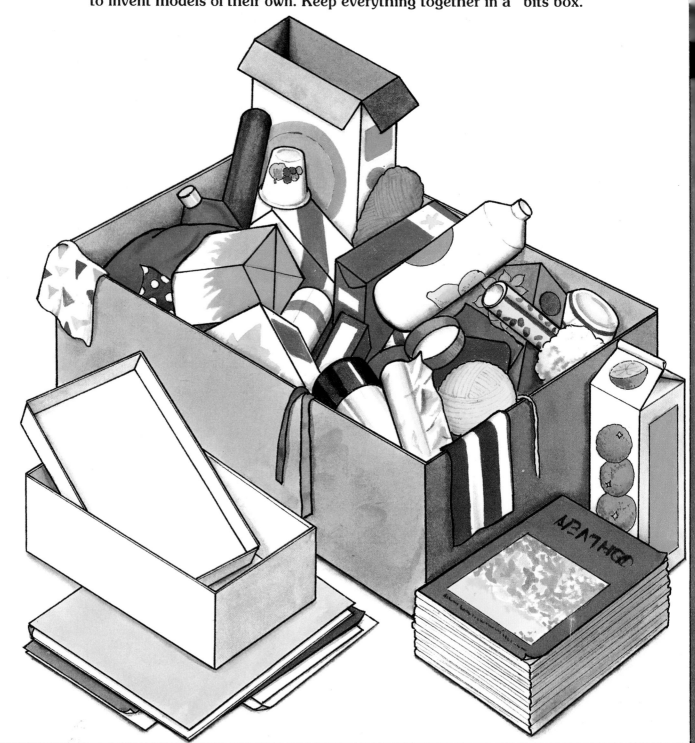

PRINTED IN BELGIUM BY

INTERNATIONAL BOOK PRODUCTION